THE BASICS OF

LINE BALANCING AND JIT KITTING

THE BASICS OF

LINE BALANCING AND JIT KITTING

Beverly Townsend

CRC Press
Taylor & Francis Group
Boca Raton London New York

CRC Press is an imprint of the
Taylor & Francis Group, an **informa** business

A PRODUCTIVITY PRESS BOOK

CRC Press
Taylor & Francis Group
6000 Broken Sound Parkway NW, Suite 300
Boca Raton, FL 33487-2742

Printed in the United States of America on acid-free paper
Version Date: 20120229

International Standard Book Number: 978-1-4398-8237-5 (Paperback)

Visit the Taylor & Francis Web site at
http://www.taylorandfrancis.com

and the CRC Press Web site at
http://www.crcpress.com

Contents

Preface

Lean is the vehicle used to achieve productivity goals, produce goods and/or services faster, with better quality and at cheaper costs. Lean is the survival of our businesses. In today's marketplace, we compete with other factories and companies from all over the world, and even in our own backyards. Your sister plants and affiliated sites are in competition for your business!

Lean does not mean eliminating a qualified, experienced, and talented workforce. It means cutting out the waste (excessive costs) that facilities have built into their processes over time. We have to compete by reducing the cost of making our products and/or services, getting them into the market faster, and at a higher quality. That seems like a daunting task, but it's only because you don't know how to see your problems yet. This book will aid you in seeing the wastes, what to do once you have identified them, and how to correct them to make your facility more efficient.

Lean is not something one person or team can do alone. It is not a program you run alongside your business, it is *how* you run your business. Lean is a common-sense approach that is used in many industries around the globe, not just manufacturing. It is truly a journey, and you cannot expect to achieve it overnight. Remember that if you are not moving forward, you are standing still and that really means you are moving backward.

Most facilities try to improve their processes by focusing on the wrong things. This book will help you to focus on the right areas that need improvement and how to correct those issues.

Over the years I have witnessed how Lean sustains businesses for growth, saves jobs, and makes benchmark sites. The fact that you are reading this book is a sign you have accepted the challenge to move forward and become a competitive supplier of goods. For that you are to be commended. I wish you success on your Lean journey, and always remember, all we have in this world is time. Once it is gone, it is gone forever.

Acknowledgments

I would like to thank those who have shaped my experiences, enriched my life, and believed in the Lean process, and in me.

To the Lean advocates in my life who encouraged, challenged and believed in me, the team in C-ville (you know who you are), J. Lovett, K. Simmons, W. Coomer, M. Gooch, and R. Elwell, I appreciate all you have done and continue to do.

To my family and dear friends for their undying support, encouragement, and belief in me and my passion, thank you.

To those who have the undying passion for Lean, as I do, relentless pursuit is the key.

About the Author

Beverly Townsend is a Lean manufacturing manager and artist from Kentucky. She is an expert in Lean, with years of experience starting at Toyota. She also holds chairman and president's awards for her Lean achievements in operation excellence for JIT kitting and waste elimination. Her experience involves teaching and mentoring in Lean as well as implementation. Her efforts have transformed facilities into benchmark plants, as well as sustained business and growth.

1

Where Are My Hidden Costs?

Most companies struggle with improving productivity, producing products or services right the first time, and meeting on-time delivery. Ultimately your goals are to reduce costs, meet your daily requirements, reduce lead time, and sustain those gains for the future of your business and growth. Not correcting issues in these areas increases your costs and will ultimately cost you and your customers.

Hidden costs are everywhere in the workforce and can be easily identified by the trained eye. All it takes is observation of the actual work. Yes, that means you must go the work area and watch, observe, listen, and see (Figure 1.1). This is known as gemba.

Gemba is a Japanese term meaning "the actual place" or "the real place." In order to understand what work you are observing you must understand what work content adds value and what does not. The work that does not add value is what you must focus on in order to improve your process. But, first things first. You really need to understand what value-added and non-value-added work means.

Value-added work content is when you change the fit, form, or function of a product—when you physically change the product or services to what the customer is *willing* to pay for. You only add value when you are doing something to transform products or services into what your customer demands as the final result. This is the work that actually makes you money!

Non-value-added work content (or waste) is anything else you do. Waste happens when you are not physically doing something that the customer demands—when your eight-hour day is ticking and you are not changing fit, form, or function. In other words, the customer is willing to pay you to physically change something or do something, providing a completed final product that they want. Anything else you do in addition to those steps is the work that costs you money and time!

FIGURE 1.1
How to locate costs.

Everything you do to provide products and services should only be what the customer wants from you every minute of every day. The customer doesn't care how you get the material where it needs to be, how you get the tools you need to work with, how long you have to wait for things, or how much money or time all that takes. They only care that they get a good quality product or service at a reasonable price, delivered on time. All the other work in your processes adds costs to the business.

Customers are going to pay only what they are willing. Wastes add to your costs and reduce your competitive edge. You must focus on the waste to remain in the game and obtain growth.

That sounds simple enough, but now you need to know how to "see" what really adds value and what does not. You have to understand what the non-value-added work or wastes are in order to correct them. There are seven types of wastes (non-value-added work content) that can be identified in the things that we do every day. These wastes (or *muda*) cause inefficiency and add unnecessary costs to your processes. These wastes are listed here in an acronym to help you remember them.

Seven Wastes (TIMWOOD)
Travel time (gathering or delivery of material or parts)
Inventory (in excess and not just in time)
Motion (of machine, team members, or material delivery)
Waiting (for materials, tools, the next unit, etc. ....)
Overprocessing (doing more than necessary to get the job done)
Overproduction (producing more than necessary not just in time)
Defects (bad quality)

Let's say that you own a business making gift boxes in your garage and you employ four people to make these boxes. There are things that they do that are "wasteful" because that is the way the processes are set up. You are paying these employees to work eight hours a day, but they are not *constantly* changing that cardboard and ribbon into a final gift box to be sold.

You expect good quality to sell the units right away to make money to pay for running your business, and to make a profit.

You need to identify these wastes in the employee work content, and balance out the work to create an even production flow to increase productivity, improve leadtime, and get the products through the processes faster for the customer. Let's explore these wastes in detail.

TRAVEL

The time materials spend moving from one place to the next while not in the process of being built or used (changing fit, form, or function) adds waste (Figure 1.2). Think of it this way: you have spent money for the material, you possibly did some work to the parts and paid manpower to manipulate them (some of that may add some value), you have set the material to the side until it can be picked up, and eventually you will deliver the material where it will sit until you are ready to use that material.

Your investment in relocating parts and supplies will not show an immediate return. Finding a way to move processes closer together, thereby reducing inventory, is the fastest return on your investment. The goal is

FIGURE 1.2
The waste in travel of inventory.

to sell your product as soon as you make it, get back the money you spent on the parts and labor, and make a profit right away. Any time you see parts or products sitting or being moved around, think of it as a big pile of money.

INVENTORY

I have heard it a thousand times, "I need all of this inventory to keep my processes running!" And every time I ask, "Do you ever deviate from your daily schedule during a shift or do you have downtime due to parts shortages? If so, is excessive inventory really fixing your problems or are you using it as a crutch? Is the inventory covering up your issues and not forcing you to fix the problems?" (Figure 1.3.)

Do you have a bottleneck today? If so, is that area running only what you need exactly when you need it? Think about it this way, are you using available capacity at work making parts that you don't need at that very moment and affecting other work areas because what you do need has not been made yet?

There is an old analogy of a boat and water that is very true (Figure 1.4). The water represents your inventory level, and your processes are represented by the boat. The rocks (or issues) are covered up by the excessive

FIGURE 1.3
The waste in excessive inventory.

FIGURE 1.4
The old analogy of plant issues and inventory levels through a boat, water level, and rocks.

inventory (water) and you are not forced to fix those issues. Instead of dealing with the correct inventory levels you need, you just build something else because you have enough inventory for that. But is that really satisfying your customers? Are you building things you don't need right now? Are you missing delivery dates? Chances are you have been "fire-fighting" because of these problems. Sound familiar?

If you lower the water level (inventory) and force yourself to build only what you need when you need it (just in time) your boat will hit the rocks (problems). Now you are forced to deal with the issues and fix them. This step is important. Facilities have run for years and never solved their problems because they are afraid to reduce inventory. The inventory is their cushion (just in case).

It has been said that excessive inventory gives the factory flexibility to run products when they don't have the other parts they need. To that I ask, "Why do you need this type of flexibility? Is it because your processes are broken? Is it because you have issues or problems? Are you fixing those issues today?"

You have spent money for material and manpower to gather the material, deliver the material, and often to manipulate the parts to what the process needs. Now the parts are sitting on a shelf waiting to be used. Meanwhile, you are building something else that will sit on a shelf. Are you getting an immediate return? Is that material immediately going to a process to change fit, form, or function? Are other areas of your facility at a standstill or building products you don't need right now? This is adding to your costs and hurting productivity, quality, and on-time delivery.

Inventory is one of the worst and most common wastes and I will discuss how to assist with these problems in Chapter 9, JIT Kitting.

MOTION

If team members are not standing at the workstation *constantly* changing the fit, form, or function, they are adding waste to the process. Locating materials, and tools, looking for paperwork, all of these things are interfering with your productivity, delaying on-time delivery, and adding labor costs to your services (Figure 1.5). Reducing this waste also makes the job easier for the operator.

Motion can also be applied to the movement of machinery or equipment. This is why 5S (discussed in Chapter 4) and bringing the parts to the point of use (POU) for the operator are so important. You are wasting money when your employees are spending any of their day walking around and locating things. This type of work content is increasing productivity.

FIGURE 1.5
The waste of motion carrying inventory to numerous places.

WAITING

Anytime a process, service, or team member is waiting on materials, product, quality checks, paperwork, another process to finish, etc., you are not changing fit, form, or function (Figure 1.6). You are still paying for the

FIGURE 1.6
The waste of waiting.

labor and products or services, and yet work is not being performed and products are not moving to your customers. This waste chips away at your productivity, delays delivery, increases lead time, and adds unnecessary costs.

OVERPROCESSING

When you do more to a product or service than necessary, you are adding cost to the product and delaying delivery time (Figure 1.7). You need to do

FIGURE 1.7
The waste of overprocessing.

only what is *absolutely* necessary to keep the product moving to the next process, out the door, and making money.

This step also applies to inventory. You should not build more parts than absolutely necessary to make the product the customer wants you to make that day. Overprocessing steals your capacity for making products the customer needs. Time lost can never be regained. It is lost forever. Overtime should not be your solution for inefficiencies.

OVERPRODUCTION

Creating more than necessary adds to high levels of inventory and steals your capacity. This can be in your warehouse, storage, or line-side inventory (Figure 1.8). If you are storing up inventory, remember that you have spent money for the materials, manpower for someone to do something to those materials, and transportation (travel) of the parts. Not to mention, if the customer drops orders or your demand changes, what do you do with

FIGURE 1.8
The waste of overproduction.

your current product that is in stock and your workforce if you don't have demand for it? You have spent money and now you can't sell your product to gain that investment back.

DEFECTS

This is the absolute worst of the seven wastes (Figure 1.9). Money has been spent to build a product once and now you are paying for it again! And all of those wastes I just mentioned that are built into your processes, you are paying for all of them again: travel, inventory, motion, waiting, overprocessing, and overproduction.

Getting your products and services right the first time is very important. Not to mention that introducing defective products into production interrupts your flow because you are reworking product and that hinders your ability to perform the service on the next item in line when it is time. This steals your capacity and delays meeting your customer demand.

Now that you understand what the seven wastes are and how they add costs, slow down productivity, and interrupt customer demand, let's discuss how you can meet customer demand efficiently.

FIGURE 1.9
The waste of defective products.

Think of it this way, there are only so many minutes in a day (or shift) that you have to complete what the customer wants you to. Do you know what that really means at the workstation where the value-added work takes place?

2

Understanding Your Rhythm

Your processes have a heartbeat or a rhythm you need to operate to in order to meet customer demand. This rhythm is called *takt time* (Figure 2.1). Takt time is defined as the maximum time at each workstation needed per unit to produce a product in order to meet customer demand every day. Takt time can be translated for every type of environment, not just in manufacturing. If you provide a service of any type to a customer, you have a takt time.

Takt time is based on a calculation using available time in a day and customer demand:

$$\text{Available seconds} \div \text{Customer demand} = \text{Takt time}.$$

Available seconds are the seconds in a shift when the facility is running and employees are expected to be building products or performing services. You should not include breaks, team communications, or cleanup (5S in Chapter 4). No one is expected to create value-added work during these times.

Example:

An 8-hour shift = 480 minutes or 28,800 seconds (one shift) − 20 minutes or 1200 seconds (two 10-minute breaks) − 10 minutes or 600 seconds (cleanup/5S time) = 450 minutes or 27,000 seconds available in a shift.

Customer demand is the number of units or products that your customer wants you to build in a shift. However, customers do not order the same number of units every day. You have to level the load (also known as the Japanese term *heijunka*), equalizing the monthly demand over the days in a month. For some facilities you may have to look back through history and get an average if demand fluctuates or is seasonal. Also note that if demand fluctuates or is seasonal, you may have to determine time based on two or three shifts.

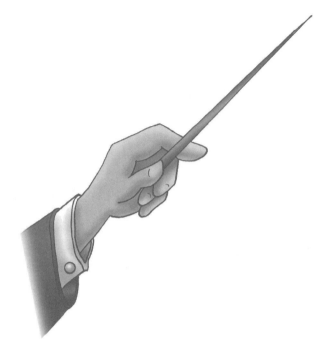

FIGURE 2.1
The origin of takt time and the orchestra leader's baton.

Example:

> Monthly demand is 9000 units ÷ 20 working days in a month =
> 450 units per shift.

So

> 27,000 seconds available per shift ÷ 450 units per shift =
> 60 seconds per unit.

Thus, the takt time is 60 seconds.

What does this really mean to you? Think of it this way: your processes are a pipe from the beginning (when the order starts) to the end (when the order is complete). The product or services moving through these processes are like water. Each process must perform its necessary tasks in takt time in order to keep the water flowing out the end of the pipe at the same pace it entered the pipe (Figure 2.2).

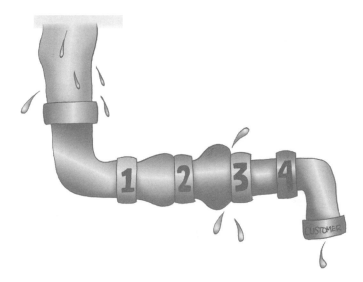

FIGURE 2.2
Bottlenecks and how constraints stop the flow of meeting customer demand.

Example:

Each station (person's work) is a section of the pipe that has a valve the worker controls (opens and closes). The valves must all open in sequence to keep the water flowing and provide the correct amount of water at the end of pipe. If they do, you will always meet customer demand at the end. Any time one of the processes takes longer than the others (the valve is not opening on time) your pipe will "clog" and the water will stop flowing. Your customer demand cannot and will not be met at the end.

For example, if station 3 has work content that takes longer to perform than the needed takt time, the next process (station 4) will wait for product (waiting) and the processes before it (stations 1 and 2) will complete their work and wait or overproduce to keep busy, causing you to miss customer demand. Numerous things can cause "clogs" in the pipe. This happens when your work content (cycle time) at any station is greater than your takt time.

Takt time is the amount of time you need for each process or station to complete its work to one unit in order to meet customer demand. Cycle time is the amount of time a process takes to complete the work for one unit. These two items are often confused. Just remember that takt time is the heartbeat or rhythm you need to achieve in order to meet customer demand. If the cycle time is greater than the takt time it is typically because of the seven wastes built into the process. The time it takes to complete these wastes increases the cycle time and causes delays in meeting customer demand.

FIGURE 2.3
A push production system and the waste it causes.

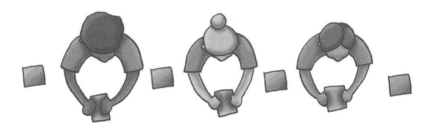

FIGURE 2.4
A pull production system and the benefits of efficiency in this type of system.

When one workstation produces faster than the next workstation, units or parts build up in your flow. This is known as a push production system (Figure 2.3).

When all processes or stations produce at takt time (your pipe valves open up in sequence), this is known as a pull production system (Figure 2.4). The last process completes the work content and pulls from the previous workstation, and so on, and so on as you move upstream (just in time).

Having a push system adds to the inventory issues you have. At the end of the production line (pipe) you can only produce one thing at a time anyway. So how does having inventory piled up at workstations in a push system help you meet customer demand? It doesn't. Most stations or processes try to produce as much as they can because they think it indicates they are working hard and staying busy. And they are, but what they don't see is that the end result is not being performed faster or more efficiently.

Okay, so you now know how to determine your takt time at each process (or workstation). And you know what the seven wastes are and how they consume your time. But, how do you know if you are meeting your takt time at each workstation, and what do you do about it if you are not?

3

Genchi Gembutsu

Genchi gembutsu is a Japanese term meaning "go see!" Get up and get your boots on (Figure 3.1)! Observe, observe, observe. The next step in this wonderful journey of waste elimination and productivity gain is to take time studies on the work content. You must document the processes and the cycle time it takes to complete the work content for each process. Many people use a stopwatch for this, but over the years I have given up the stopwatch for a video camera. It is easy to rewind, and the time for each element of work is easily documented on the bottom of the screen. I encourage you to stay at the workstation during recording however. There is nothing like being there at the real place the work is being done (*gemba*) to understand the work content and how the processes interact and link to one another.

Pick one unit or service to observe at a time. You must video your team members doing their jobs and document what they do and how long it takes them to do it. Start at the beginning of the process and follow it all the way to the end. Record everything and *do not* stop the camera.

Beware that some team members will get wise to this technique and will move more slowly or not work to their potential trying to make their job seem more time consuming. On the other hand, you may have team members who want to overachieve, but they cannot keep up that pace all day long. When you begin to videotape, try to select team members who work at a good even pace (not too fast and not too slow).

I suggest that if you are not familiar with all the processes you should have the workers narrate what they are doing. They do not need to stop and look at the camera to explain (that increases their time), all they need to do is talk as they work.

Now let's say you have video documented every job in the process and the workers have narrated what they are doing. It is time to formally document these steps and evaluate the work content (Figure 3.2).

FIGURE 3.1
The idea of *genchi gembutsu*, "get your boots on."

FIGURE 3.2
The tools needed to do a time study.

You will need the following:

- your camcorder with all the data
- a quiet place to sit to review all of the data
- a tablet of lined paper
- a pen or pencil

- a red marker
- a green marker
- a calculator

Document the small incremental steps the operators take, with the cycle times from the video included. The smaller the increments of work you document the better. This way you can easily identify the non-value-added work content of the job so that you will know how to assist with reducing the cycle time and thereby make your processes more efficient. Remember the pipe and takt time (Figure 3.3)? The data you have gathered will allow you to see when the processes cannot meet the takt time and when you are unable to meet customer demand.

On the left-hand side at the top of the lined paper write the name of the work area (Line 1) and the type of unit or process you recorded (Product A) write the workstation name or number (Station 3). Then, on the top right-hand corner write the word Time.

Rewind the videotape and start with the first workstation. It is very important to break down these steps as small as you can. Always separate value-added from non-value-added work content (Figure 3.4). You need to be specific. This becomes very important when you need to rebalance work content. If you don't know what the part is they are working with, or how long it takes to do each step, you won't know what work content can be moved to another station or processed later.

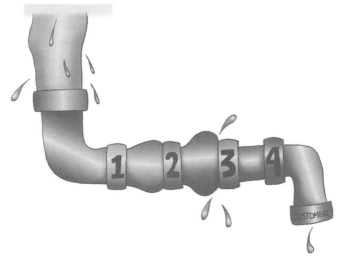

FIGURE 3.3
Reinforcing the flow, bottlenecks, and constraints.

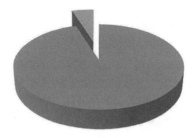

FIGURE 3.4
Graph of a typical value-added and non-value-added work content process.

Example:

Line 1, Product A

Station 3	Time
Step 1: Find and pick up flat cardboard 0:00–1:15	75 seconds
Step 2: Find and pick up tape dispenser 1:15–2:15	60 seconds
Step 3: Fold cardboard box and tape 2:15–2:36	21 seconds
Step 4: Return tape dispenser and send the box to the next station 2:36–4:00	84 seconds

Total cycle time for Station 3 = 240 seconds

The sum of these steps is how long it takes station 3 to get the job done (station's cycle time). The sum of all the stations is how long it takes the unit or process to be completed (or the overall cycle time for the whole unit). The total time for all stations is the production line's lead time, that is, how long it takes to complete a unit and send it out the door to the customer. Understanding this will also assist with the scheduling process. Scheduling and customer service rely heavily on understanding lead time. Reducing lead time improves customer satisfaction. (Inventory in queue is a factor in lead time as well — waste.)

You need to record these work steps three times to get an average time of the work performed. The operator will usually never perform the work content the same time every time. Analyze your three time studies, and if the work content does not have a difference of 10%, then take the repeated time for each step. If there is not a repeated time, take an average of the three times for that work step. If the times differ by more than 10%, record the work content three more times and repeat the steps until you find a repeated time or take an average.

Once you have written down all of the steps of the job, it is time to categorize this work content into value-added versus non-value-added work. To do this, simply start at the first step and highlight the cycle times. (Red for non-value-added work content and green for value-added work content.) Red indicates that something is wrong (just like stop lights, stop signs, etc.), and green indicates things are okay.

Example:

Line 1, Product A

Station 3	**Time**
Step 1: Find and pick up flat cardboard 0:00–1:15	NVA 75 seconds
Step 2: Find and pick up tape dispenser 1:15–2:15	NVA 60 seconds
Step 3: Fold cardboard box and tape 2:15–2:36	VA 21 seconds
Step 4: Return tape dispenser and send box to the next station 2:36–4:00	NVA 84 seconds

Total cycle time for Station 3 = 240 seconds

Non-value-added (NVA) work content = 219 seconds

Value-added (VA) work content = 21 seconds

Your processes have developed this way over time. These wastes have become "normal" and are tolerated or even accepted in our lead times to our customers! You have set up and established jobs that are inefficient and it makes the work harder on the team members.

If you can reduce your non-value-added work content, then the faster you will be able to please the customer.

These cycle time steps need to be documented for every workstation. This is tedious work. But remember, once created, it does not need to be re-created until you have an engineering change, reduce the waste, or have new product.

Once you have completed this process for one job or workstation, if you see more red than green that tells you that you have an inefficient job or workstation that is full of waste. Most work content in your processes

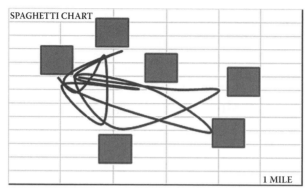

FIGURE 3.5

A spaghetti chart and the waste of motion in processes and how to visually identify the motion.

will only be 5% value added! Many companies try to improve their value-added work because they do not understand the gains they will have if they focus on the non-value-added work content.

The next step on this journey is determining the walk path (motion) each operator uses to do his or her job. The Lean tool used for this is a spaghetti chart. These charts are an aerial view of the workstation (to scale is not necessary) and the walk path the operator takes. The distance is recorded on the chart using a pedometer. They are called spaghetti charts because they end up looking like a bowl of spaghetti when completed (Figure 3.5).

To draw a diagram of the work area from a bird's eye view, rewind the video and trace the path the worker takes to get the job done. Go to the floor (*genchi gembutsu*) and walk a mile (or 20) in their shoes.

These charts will help you to realize the amount of motion waste in the process. This tool is also used to help you see the time it takes the operator to gather tools, parts, and materials and then work. To be the most efficient you want the operator to stand in one spot and add value. You need to focus on cutting out the waste. There is more on eliminating this waste in Chapter 9.

A standard work combination sheet can also be used to identify the steps and waste in a process (Figure 3.6). This chart lists the steps of the operation according to the times you have collected and visually displays them to easily identify work content waste. In this example, walking (the curved line) shows the time the operator walks and assists you in seeing opportunities to improve material handling.

You have the cycle time with value-added and non-value-added work content as well as the distance (motion) in the process. It is time to learn about ways to reduce wastes.

Standard Work Combination Sheet:

Part/Model Name: BOX A		Department: LINE 1		
PROCESS: CARDBOARD PREP		Operator: JOHN DOE	TAKT Time: 60 SECONDS	Date: 1 / 1 / 11

Step No.	Process Step Description	Time V/A	Time NVA	CYCLE TIME (In Seconds)
1	GATHER CARDBOARD FROM WAREHOUSE		25	
2	PLACE CARDBOARD IN CUTTING MACHINE		5	
3	PROGRAM MACHINE FOR CUTTING SIZE		5	
4	PUSH START		5	
5	MACHINE COMPLETES CYCLE (MACHINE TIME)	15		
6	MACHINE CUTS CARDBOARD (MACHINE TIME)		5	
7	REMOVE CARDBOARD FROM MACHINE		5	
8	WALK CARDBOARD TO LINE 2		30	
	TOTAL	15	80	

TOTAL CYCLE TIME: 95 SEC

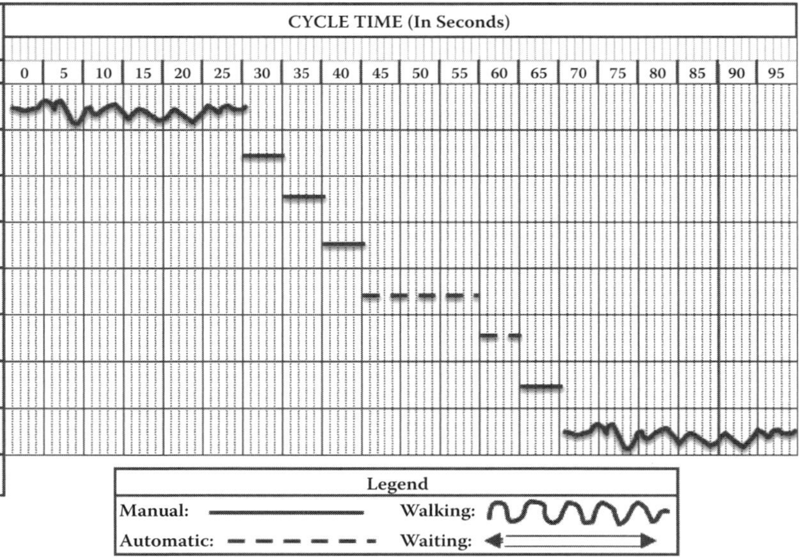

CYCLE TIME (In Seconds): 0 5 10 15 20 25 30 35 40 45 50 55 60 65 70 75 80 85 90 95

Legend
Manual: ——— Walking: ∿∿∿
Automatic: - - - - Waiting: ←→

FIGURE 3.6
A standard work combination sheet.

4

5S

5S is a series of five words all starting with the letter S describing a process of an organization (Figure 4.1). It is a philosophy for keeping workstations organized and clean. Many facilities state that the majority of their issues are caused by poor 5S procedures. 5S is a discipline and foundation for any Lean facility.

5S
Sort
Set
Shine
Standardize
Sustain

Let's explore these in more detail.

SORT

"When in Doubt, Throw It Out."

Clutter in your workstations leads to waste in your processes. Searching through tool bins and disorganized parts racks increases cycle time and reduces productivity, quality, and on-time delivery while increasing costs.

Sort is the first step in 5S (Figure 4.2). You must eliminate all things that are not needed for the workstation. If you are not certain if an item is needed (doubt), discard it. Sorting at a workstation is typically completed with the leaders, Lean efficiency personnel, and of course the operator.

FIGURE 4.1
An illustration of 5S.

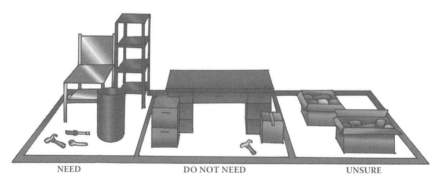

| NEED | DO NOT NEED | UNSURE |

FIGURE 4.2
5S and sorting objects at a workstation into separate categories of need, do not need, and unsure.

For the Sort three locations need to be marked off on the floor to sort your items into piles:

1. Needed items
2. Unsure if needed (store away from the process for a month and reevaluate)
3. Do not need (these items should be placed in the red-tag area)

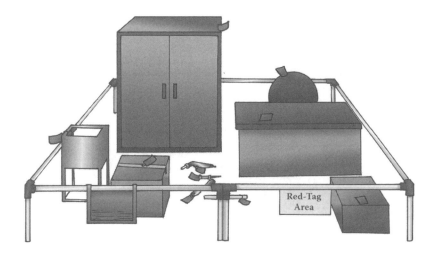

FIGURE 4.3
A red-tag area where unneeded items are properly disposed.

The red-tag area is a location in the facility for discarded items (Figure 4.3). A red tag is placed on the item to explain why it is located there (not needed, broken, etc.). Items that are not broken are then available for other work areas that may need them. This prevents purchasing items that you may already have available. Periodically the red-tag area needs to be cleaned and unused items disposed of.

SET

"A Place for Everything and Everything in Its Place."

The second S is very important in reducing the cycle time at each workstation to improve productivity, quality, and on-time delivery. Your time studies have shown you the waste that each process has in gathering parts and tools to complete the job. The spaghetti chart and standard work combination sheet are also very telling of this waste. In order to reduce the operators' cycle time, you must place what the operators need to perform their daily tasks efficiently at their fingertips (point of use).

Shadow boards contain the workstation's tools with an outline (shadow) of each tool (Figure 4.4). The tools are placed on the board in the sequence in which they are used during the process. The shadows ensure that the tools are placed back in the same location after each use. This creates a habit

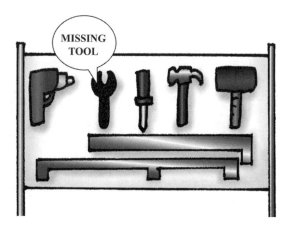

FIGURE 4.4
A shadow board and its purpose in 5S.

for the operator and organizes the workstation, reducing the cycle time for locating tools.

Designated locations on the floor for everything are an example of visual control for the leadership team as well as organization for the operator. Any items that are not located in their designated area signal an issue and are easily detected in a matter of seconds. Remember, if products or parts are not in the process of being changed in fit, form, or function, then there is a problem and those parts are not making money. Those parts are costing you money with no immediate return.

Work in process (WIP) is material that is in the process of going to the next workstation. Some levels of WIP are acceptable and are needed for line spacing between processes. An example of this is before or after a machine changeover. Just keep in mind that the longer a material is sitting idle, the longer your return on that investment, and the more capacity you have wasted working on parts you don't need right away. Think just in time.

Industrial tape is preferred when setting locations on the floor for equipment, trash cans, parts racks, shadow boards, etc. Painting lines indicates that you do not plan to move (or improve) these areas in the future and you have reached a level of perfection. Through my experience I have learned that sweeping and then mopping the floor before applying the tape will ensure that the tape will last for many months, or even years. Rubbing the tape into the surface aids in adhesion as well.

FIGURE 4.5
Shine, keeping areas clean.

SHINE

"Keep It Clean."

Once a sort and set are performed, keeping the area clean becomes an easier task (Figure 4.5). There is less clutter, and the area is now organized for the operator. Team members are responsible for shining their work areas daily. Shine (sweeping areas daily and wiping down racks and equipment) has numerous benefits:

- Sustaining waste reduction improvements.
- Visitors get a feeling that you have control of your processes and you are capable of producing a good product.
- Leaks and issues with equipment are easily spotted daily and improvement can be made to prevent equipment downtime.

STANDARDIZE

"This Is What Is Expected Every Day in Every Area."

To meet the expectation the third S, "shine," needs to be documented on a 5S checklist that the operators fill out daily to standardize the process. The checklist should include all the daily 5S activities to be performed. This checklist is verified for operator participation by the workstation supervisors and must become a job requirement to maintain your waste elimination and cycle time reduction efforts. Each day of the week is listed with activities to be completed by the operator. Some activities may be placed on a weekly schedule, including wiping down lighting, mopping floors, replacing damaged tape, etc.

A pilot cell for 5S is recommended. Once the area has been through a sort and set, maintaining the shine requires a culture change. Any new process must have procedures documented to maintain the process.

Total preventative maintenance (TPM) is a process of daily checking of equipment and tools to ensure that they are fully operational for the next shift. Equipment downtime will cause issues with productivity, quality, and on-time delivery. Maintenance schedules are often established through a process from the maintenance department. However, 5S checklists should incorporate simple TPM checks for the operators to verify that all machinery is working properly. Some simple tasks may be assigned such as lubrication of equipment and small replacements. This is known as autonomous maintenance. If an issue is found on the daily checks, the line support must be notified immediately to prevent downtime.

SUSTAIN

"Audit the Process to Ensure Sustainability."

Any new process requires audits to ensure that it will sustain itself (Figure 4.6). Old habits are hard to break, and constant follow-up is necessary for a new program's success. Leadership and management need a process to audit the 5S checklists and ensure the actual work is being completed by the operators to maintain 5S over time. Audits of the checklists should be developed and completed weekly or monthly my upper management (Figure 4.7).

The results of the audit should be posted at the workstation to visually inform the team of their 5S success and opportunities for improvement. Action items for noncompliance to the standard should be posted in the area and checked off when items are completed.

5S AUDIT

If 100% compliant, question is worth 2 points.

If partially compliant, question is worth 1 point.

If not compliant, question is worth 0 points.

MONTH: JUNE	SCORES
SORT	
All unneeded items are removed from the area	2
All trash cans have been emptied	2
SET	
Area does not have any items out of place	1
All tools are present with shadows	2
All "Set" tape is clearly marked and clean	1
SHINE	
Area is swept clean	2
Area is mopped	1
All machines are clean	2
All lighting is clean	2
STANDARDIZE	
Every work station participates in 5S	2
Every work station maintains 5S daily	1
SUSTAIN	
Area has improved 5S scores month to month	1
TOTAL MONTHLY SCORE	19

* *MAX POINTS = 24*

FIGURE 4.6

An audit form to sustain your 5S efforts.

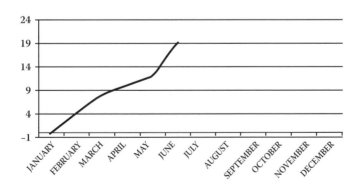

FIGURE 4.7

A graph of the sustainability of 5S for a workstation.

Many facilities provide quarterly rewards for the most improved workstations in 5S to encourage healthy competition and to recognize good work. After all, sustaining this process improves your productivity, quality, and on-time delivery. 5S is very important to your success.

5

Line Balance and Yamazumi

Now that you understand the actual cycle time for each workstation and its work content, it is now time to complete a cycle time chart. (These are typically posted next to the operators' work cells.). This chart will graphically show you the amount of time it takes for each process and when your processes do not allow sufficient time to meet customer demand (Figure 5.1).

The bottom of the chart lists the workstations or processes that you currently have in your work cell. Station 1 goes to the left and the last station to the right. Each bar of the chart represents the cycle time for the job. This is for each operator's work content on one unit. The time line is represented on the right-hand side of the chart, with the takt time line drawn in.

Each process that is over the takt time is a process that cannot meet customer demand. Likewise, the processes that are under the takt time represent waiting in our work flow. Remember the water and the pipe illustration?

Now, you can visually see that your work content is not balanced evenly and how long it takes to complete a unit. (The addition of the cycle times at each station is your overall cycle time.) For immediate improvements you want to focus on only the processes that are over the takt time.

However, this chart is not detailed enough to assist you with seeing waste or how to balance the small elements of work to keep your pipe flowing. For that you will need to create a yamazumi chart (Figures 5.2 and 5.3).

Yamazumi is a Japanese word that means "to pile into heaps or stack up." You will "pile" your work content into heaps to represent the work content at the workstation.

The yamazumi chart is designed to help you rebalance work content so that you can meet customer demand and keep your pipe flowing. You must rebalance the work at each station to get your processes under or takt time in order to meet customer demand.

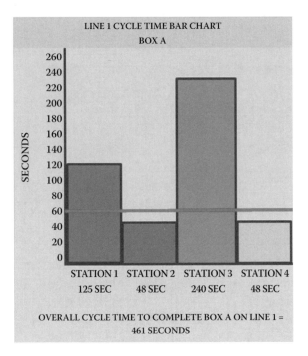

FIGURE 5.1
A cycle time bar chart.

FIGURE 5.2
Balancing work content on a yamazumi chart.

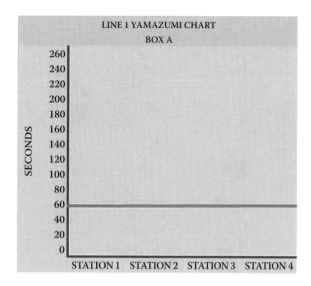

FIGURE 5.3
How to create the blank form for a yamazumi chart.

The top of the yamazumi chart lists the process area you are currently documenting and the product you have chosen to document. The left-hand side of the chart contains the time line. Every product line in your facility will probably have a different time line based on work content. Some areas will require using minutes and others may require using seconds.

For example, one area of the plant may have a 20-minute takt time and another may have a 57-second takt time. Remember, this is based solely on customer demand. The bottom of the chart lists the workstations or processes that you currently have. Station 1 goes to the left and the last station goes to the right.

Next, you will use a text box to represent *every* step at the workstation. You will type the step description and time it takes to do this step in the text box. Remember to keep your work elements small while doing your time studies. You need to completely separate non-value-added work content from value-added work content. Then, color the text box red for non-value-added work content and green for value-added work content (just like you colored it on your lined paper) (Figure 5.4).

This is continued throughout the process and the blocks are stacked on top of each other until you have a pile of work content that equals the total cycle time for that operation. These steps are followed until every process in the work cell is completed. Now you have the steps it takes to build a unit and the work content at each station listed. The total cycle time for each workstation is listed underneath the title of the station.

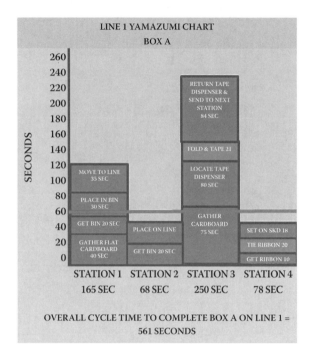

FIGURE 5.4

A completed yamazumi chart with the value-added and non-value-added work content and unbalanced workstations.

Now you need to draw a red line on the yamazumi chart where your takt time is. This way, you can see which stations are over the takt time and which are under the takt time. Remember, those that are under the takt time contain the waste of waiting.

The differences in the cycle times for each station will be very clear. Over a period of time you may have moved work content to the fastest worker and taken work content away from a slower worker. Then one day, your fastest worker leaves the workstation and another worker is brought in, but the new worker is not as fast as the previous worker. You move work content again, and so on, and so on. You end up in a vicious cycle until your work content is not balanced and your quality, productivity, on-time delivery, and costs suffer.

Hopefully you are beginning to truly understand what the work content is and how processes really function (or don't function). So, what do you do with all this information? The first thing is to analyze your spaghetti charts, standard work combination sheets, and yamazumi charts and start to think about cutting out the waste built into your processes.

VARIABLE WORK CONTENT

Variable work content is the work content that is not repeated every time or every cycle of a process (Figure 5.5). An example of this may be options on certain units that run down a line with units that do not have options. Typically when you do not have variable work on the next unit, there is time waiting for the next unit that will have it. Variation in your process does lead to waste and should be evaluated carefully in each value stream. The weighted average of the work content on the line is based on the work content for all processes. This average will change as you reduce the time of all wastes.

If the overall cycle time of two units has a 30% or greater difference, those units do not belong in the same value stream because of this waste, causing bottlenecks in processes and interruptions in product flow.

What are the wastes in each process that you can reduce or eliminate based on improvements you can make using 5S, bringing parts and tools to the operators (point of use), reducing inventory, or holding a single minute exchange of dies (SMED) event?

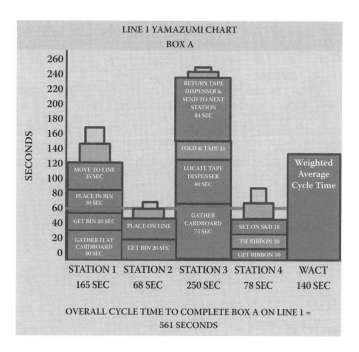

FIGURE 5.5
A completed yamazumi chart with variable work content.

SMED is the concept of reducing change at a workstation. Any time you change over dies or tools at the workstation, you are not changing fit, form, or function. You are not changing the product into what the customer is willing to pay for. The idea is to reduce the time it takes to change out tools and dies to keep the product moving to the next operation and out the door to the customer. Arranging tools and dies (5S) and placing them at the point of use of the operator will assist in maintaining a value-added process.

An example of SMED is race cars and pit crews. They practice SMED and understand their wastes in the time it takes for a pit stop change in order to stay in the race. They, too, reduce waste through time study observation to remain competitive.

6

Rebalance

Examine the information of the wastes in your process and make a list of all the time you can save at each step. Write down your improvement ideas for future projects. Simulation of the new work content may be necessary to understand how much waste you have eliminated and how much time the step will take now. Not all non-value-added waste can be eliminated. That does not mean that these steps now add value, but it is considered necessary waste. Make sure you challenge the old habits and culture. Question everything you think may be necessary. You will now take this new information and manipulate your red blocks on your yama-zumi charts to reduce your non-value-added work content and decrease your workstation's waste.

NOTE: Variable work content is not listed here, but don't forget to add it to your charts.

Overall cycle time (the total time for all workstations to complete the work for one unit) will also be reduced as you eliminate waste and increase throughput in your work cell.

Now that you have evaluated and eliminated the waste on paper (reducing or eliminating the red blocks) it is time to rebalance the workload and align your processes to meet customer demand by moving work content from one workstation to another (Figure 6.2). You must keep in mind that a certain work sequence must be maintained for quality and ease of assembly.

This is why it is very important to document the content in the text blocks. Once you start moving things around you may lose track of the steps you have moved and what that work content was.

NOTE: Save the original version of your chart and manipulate a copy (Figure 6.1).

FIGURE 6.1
A balancing scale, emphasizing the balance and harmony you are trying to attain.

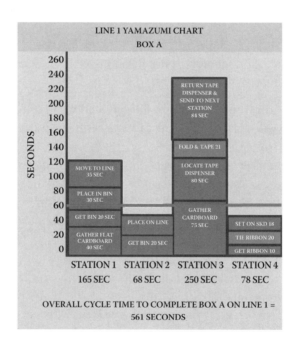

FIGURE 6.2
An unbalanced yamazumi chart.

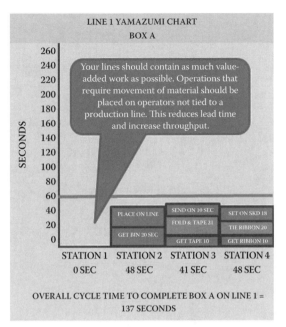

FIGURE 6.3
A balanced yamazumi chart with waste removed.

No balance will ever turn out to be perfect. Some steps take longer than others. But the goal is to get each station's cycle time as even as possible and under the takt time.

The goal is to balance the work content for each station at 95% of the takt time (Figure 6.3). You never want to balance right to the takt time line. You need some wiggle room for the "human factor." However, balancing less than 85% is accepting the waste of waiting in your processes.

LEAVING THE WASTE IN

Leaving the waste in our processes and balancing work content to meet demand suggests that we need more operators (Figure 6.4). This is not efficient and adds more costs. It is vital to your success to eliminate the waste.

REBALANCE REMOVING THE WASTE

Notice the overall cycle time reduction on the line. What a difference in lead time! You may find that when you start to rebalance, one worker may

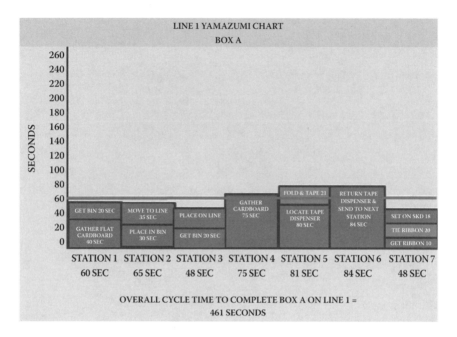

FIGURE 6.4
A balanced yamazumi chart with waste included.

need to be removed from the production line. Don't think that they will no longer be a part of your organization. As you begin to make all of your processes largely value added, the non-value-added work content typically moves to material handling.

These workers are very valuable, but they are not changing fit, form, or function. So the non-value-added work content will typically fall to these positions. Your new line is not going to use an operator to walk away from the line and gather material. They are added to a new process of kitting. We will discuss this in Chapter 9.

Periodically your takt time will change based on demand and thus the balance will change. You may need to add operators if your demand increases or remove operators if your demand decreases. We will discuss how to determine operators based on demand in Chapter 7.

Your lines should contain as much value-added work as possible. Operations that require movement of material should be assigned to operators not tied to a production line. This reduces lead time and increase throughput.

I suggest you have a leader or experienced personnel with you when you are ready to rebalance your yamazumi chart (Figure 6.5). You must

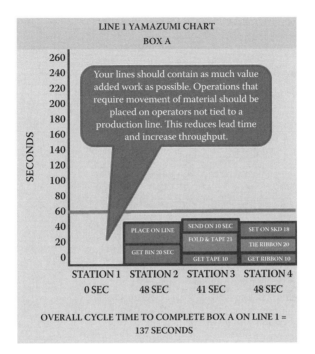

FIGURE 6.5
A balanced yamazumi chart with waste removed.

keep the steps in the sequence needed to build the product correctly. These experienced personnel know where things can and can't move to improve processes and keep the steps in sequence. But don't be afraid to challenge them either! These people have been doing their jobs the same way for so long, they tend to not think outside of the box.

Once you have a final rebalance, redraw your standard work combination sheets and spaghetti charts to match your new balance, and look at the difference. Most likely you have been operating with extra people in the production cell and with large amounts of waste. You are now on your way to a new Lean facility.

7

Flex Plans

"Now, what happens when my customer demand changes?" How do you know the manpower you need? How do you know what each person's work content will be?

To determine manpower needs you use the formula

$$\text{Overall cycle time} \div \text{Takt time} = \text{Manpower.}$$

The overall cycle time (OCT) is the sum of all the processes for one line or work cell. If you begin making a new product, you must do your time studies again and understand the actual work content. If the product stays the same, the work content will only change when you make improvements and reduce wastes (which is always wonderful). The formula for takt time is provided in chapter 2.

Example:

137 seconds (total overall cycle times) ÷ 60 seconds (takt time) = 2.283, or 3 operators.

You can use the manpower formula to assist you with demand and manpower needs. Reducing waste in your process reduces your overall cycle time and can ultimately reduce the manpower needed to meet customer demand. A yamazumi chart is used to help you document new standard work and to train on the new work content at each workstation. You can drop your takt time line, and adjust the text blocks. Remember, you can use this chart forever. Save each new rebalance, because your takt time may return to that again and then you will not have to reinvent the wheel. Now you have a tool to make sure you are on track with a Lean process.

You have balanced the work content on paper and it all looks good. It is time to give it a try. Notice that I did not say this is final. You must try it out and see how it works. Adjustments may be necessary when you take theory to the real world. I suggest that you practice your new balance for at least a week. You must stay at the work cells and make sure all your balancing ideas have worked out. That's right, *genchi gembutsu*. You can't do this at your desk.

8

Standard Work

Parts need to be located at the team members point of use and in the correct sequence in which you have determined the operator will conduct the work according to the yamazumi balance you have created. This sequence of materials will allow the team members to ensure that they follow standard work and cut down on non-value-added work.

So, you have a new yamazumi chart less waste and new work content at each station, but you don't know how to implement it. Once the new balance is finalized you need to type the new sequence of events on a standard work form. This form is for the operators' instructions and training documents. Standard work is the sequence in which the operators are required to do the job. Key points can be added to the standard work sheet, or in addition to it (Figure 8.3). Key point sheets list in detail the specifics of the step including quality and safety points.

Following this exactly every time will ensure that new processes are sustained, and it will assist the operators in forming correct working habits that will decrease their cycle time as they find their rhythm with the work content from repeatability.

Start with the first new job you have created on the yamazumi chart and type out the steps of the job in sequence starting at step one. The operating sequence of events (SOE) or standard work should contain

- Safety key points (can be on a key point sheet as well)
- Each step in detail
- Time required for each step
- Photos of steps to clarify the work required (can be on key point sheet)
- Quality key points (can be on key point sheet)

If the standard work is followed exactly and issues still remain in the process, the standard work needs to be evaluated. It is a living document

FIGURE 8.1
Illustrates opening to chapter and *genchi gembutsu.*

LINE 1	
SUPERVISOR: JOHN DOE	**SHIFT 2ND**
1 GATHER TAPE FROM DISPENSER	10 SEC
2 FOLD BOX AT EDGES AND TAPE BOX TOGETHER ALONG SIDES	21 SEC
3 ATTACH RIBBON TO BOX IN A NEAT BOW	15 SEC
4 SEND BOX TO THE NEXT STATION	10 SEC
	56 SECONDS

FIGURE 8.2
The use of time studies in standard work.

and is should be updated as operators have suggestions for improvements to the work they perform. However, only the line support and leadership should change the standard work, and until the document is changed, operators should follow each step in sequence.

It is very important that each team member understands and signs off that they agree to do the job this way every time. Repetition makes them form a habit, and over time they will become efficient at their jobs, thus improving quality. Operators are no longer trying to determine what part of the job comes next, and they can easily see mistakes passed on to them by other team members. Standard work is also a training document for other team members on job rotations for ergonomic reasons.

Standard Work

Line 1		Station 2	
Box A		Supervisor: John Doe	

No.	Elemental Description		Time
1	Gather tape from dispenser	✚ ⬤	10
2	Fold box at edges & tape box together along sides		21
3	Attach ribbon to box in a neat bow See key point sheet for detailed instructions	◆	15
4	Send box to the next station		10
	TOTAL CYCLE TIME		56

Key Points	
✚	Safety (gloves required)
◆	Quality
⬤	PPE (gloves required)

FIGURE 8.3
A standard worksheet.

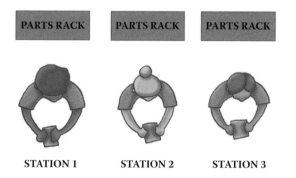

FIGURE 8.4
Workstation design.

Next, you need to create a new bird's eye view map of the assembly line so that all team members will know their place on line and their work content. Pass out these forms (yamazumi charts, standard work combination sheets, spaghetti charts, and the new standard work and/or key point sheets) to the team members at least one week in advance of the trial to give them time to absorb all this information. Also, if the work cell is new, take industrial tape and make Xs on the assembly line floor and add work-station numbers to the tape so that workers will know where they should stand when the big day comes. These things make the move much easier.

9

JIT Kitting

I hope you now see that the most common waste in facilities is the waste of gathering and moving material. You have placed materials at the point of use of the operator and moved material handling so that the team members that add value are performing less waste for every product. This has improved productivity, quality, and on-time delivery, but there is another step you can take to reduce waste even further.

Examining your new yamazumi charts and standard work you can see there is still waste in gathering material, even if it is only the operators gathering material from behind them on a material rack. Imagine if the parts are with the unit at the very moment you need them and the operator does not even have to turn around: this is just-in-time (JIT) kitting (Figure 9.1). Wow, now that is really becoming efficient!

There are other advantages to kitting besides eliminating more waste and increasing productivity. You can improve quality by having the parts available and limiting the options of numerous parts to the operator. You can ensure that the right parts are installed, thereby correcting your issues with the bill of material (BOM). When kitting all of the parts for the unit, the parts are chosen from the BOM. But we'll get to that later.

So where to start such a daunting task? Again, you start with your yamazumi charts. Notice that all the work is based on time. After all, time is all you have to make what the customer wants. Any time that is lost can never be recaptured, it is gone forever. Identify the waste in your yamazumi charts even further. Look for the opportunities that lie in the gathering of material.

You may need to simulate gathering parts that are now at the operator's fingertips to understand the time you can save. Once you have determined the new steps and the time for each workstation, include those steps on your yamazumi chart and look at the time you have saved! You guessed it,

FIGURE 9.1
The idea of a JIT kit.

STATION 1 STATION 2 STATION 3

FIGURE 9.2
Separating the work content at each station to determine kit size.

it's time to rebalance again. And these new steps are the foundation for a new updated standard work.

Since this looks so good on paper, how in the world do you make it a reality? There are several steps you must follow in sequence to make this program successful. I will now discuss these steps in detail.

KIT SPECIFICS

What Type of Kit?

You must determine the size and material handling methods for your kits. You need to locate a space in your facility to lay out parts on the floor for one unit and determine what parts are needed at each workstation. Gather your new yamazumi charts and updated standard work. If your particular line builds more than one type of unit you will need to collect all the parts for all the units, but keep them separated at this point.

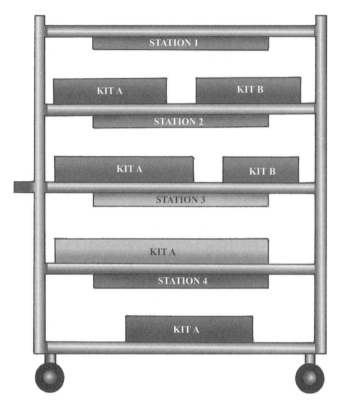

FIGURE 9.3
A one-unit kitting cart.

Now that you have all the parts for one unit, lay out the parts on the into sections representing the workstations and the work content (Figure 9.2). At this point you will determine which type of kit you should use for your line. Is it possible to put all the parts into one kit and let it go down the line with the unit? If so, this is the easiest way to do kitting. We will call this a one-unit kit (Figure 9.3). If it is not possible (because of part sizes or the spacing needed for the kit) you may need to kit each station separately. This form of kitting requires more work, but is absolutely possible. We will call this multistation kitting.

Once you determine which type of kit you will use for your line (a one-unit kit or a multistation kit), it is time to determine the kit size.

One-unit kits can be small or large. They should be able to hold all of the parts for one unit and be able to flow with the unit down your production line. The parts need to be organized and easily accessible to the operator. Many companies use small trays with cut outs to hold the parts, while other companies with larger kits use a tray or slotted cart that enables ease of use by the operator. Larger carts may require some engineering to determine how they will flow down the line without having to be pushed. After all, using manpower to push a kit is a waste.

Some kits have small tracks in the ground to guide them. Sophisticated facilities even use robots to drive kitted parts on a track across the plant to the line. If you have this option, good for you! If not, you will have to invent a way to easily move the kit on and off the line when they are empty.

Multistation kits need a little more innovation. In this scenario, each station must receive its parts specifically for each unit. Determine the size of the kits based on each station's work content and the parts you have laid out on the floor. Remember to separate the parts in some sort of tray for ease of use by the operator. Each station and kit may be a different size. For ease of material delivery, I suggest making all the kits the size of the largest kit, if possible, but always keep in mind the ergonomics of the operator and material handler. Size and weight is very important. Each kit container requires labels for each workstation and each line production cell as identification. Each workstation should also contain the corresponding identification for the production cell and workstation. This allows kits to be placed at the proper stations from the tugger and placed in the proper location for the tugger pickup in the supermarket (which I will discuss later in this chapter).

MATERIAL DELIVERY

What Type of Transportation?

Now that you have the kit size for a one-unit kit or a multistation kit, you can determine the material delivery of the kits to the line. Innovative Lean thinking suggests that no more than one hour line-side inventory should exist on your line. The purpose is to reduce inventory back to the supplier (just in time) and allows for scheduled adherence to meet customer demand. It also makes 5S a lot easier because your line will only have workstations with very few parts and no parts racks holding lots of material.

Material delivery should take the form of a tugger or kitting cart and of course a material handler (Figure 9.4). Your material handling department is already delivering parts. You are just redefining how much and how often.

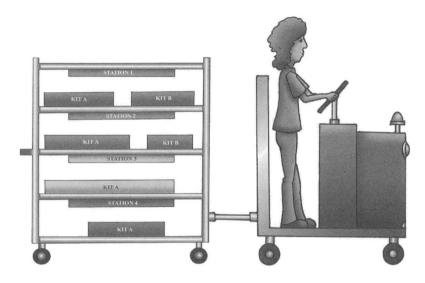

FIGURE 9.4
A tugger route with JIT kitting delivery methods.

Kitting carts are typically used for small parts for delivery to one-unit kits. Tuggers are typically used for larger multistation kits. Regardless of the type of delivery you choose, never forget proper ergonomics for the material handler and operator.

WORKSTATION DESIGN

Designed for Kits, Tools, and Ergonomics

Workstation design is based on the kit type, size, and material delivery method you have chosen (Figure 9.5). In one-unit kits you should only need a station to hold tools and hardware. Multistation kitting will need workstations to hold tools and two kits of material, one kit for the current parts the operator is working on and one in the queue for the next unit coming down the line.

Workstations should never contain more space than is absolutely necessary for parts and tools. Excess space in workstations leads to clutter, hoarding of parts, and personal items. This leads to poor 5S. Shadow boards are an important part of your workstation design (Figure 9.6). They eliminate the waste of locating tools and are part of 5S. Determine the tools needed for the job and the sequence in which the tools will be used according to your standard work.

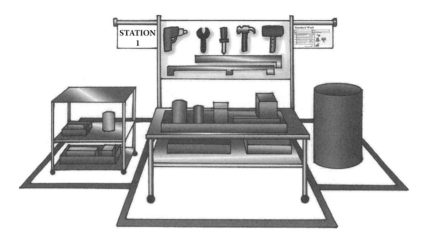

FIGURE 9.5
Workstation design after JIT kitting is determined.

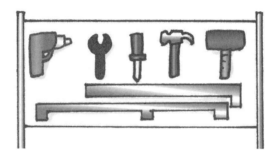

FIGURE 9.6
Shadow boards and their purpose.

The parts delivery to your workstations should be thought out carefully utilizing correct ergonomics for the operator and the material handler. The tugger cart containing the kits for the workstations should align in height with the workstation for ease of delivery.

LINE-SIDE LAYOUT

Layout for Station and Tugger Delivery

Line layout is very important to your material delivery success. You must take into consideration your workstation size, tugger or cart delivery and parking, and tugger turning radius (Figure 9.7). Although you will no

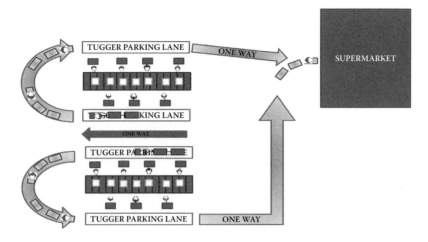

FIGURE 9.7
A tugger route and delivery path.

longer have material racks at line side, you will have tugger lanes for delivery or one-unit kit deliveries. This can be determined by re-creating the line to scale on paper.

Tugger routes need to be defined and one-way traffic for aisles needs to be determined. This will really be affected by the placement of your supermarket, where parts will be picked up and collected for delivery to the line.

SUPERMARKET DESIGN

Size and Layout of the Supermarket

So why in the world would you pull parts off the line and then just place them somewhere else in the building? The idea of a supermarket is to eventually eliminate the supermarket (Figure 9.8). This is an interim step to a final result of ordering material in smaller batches from your suppliers according to your demand. This will increase inventory turns and reduce cost. Interim steps are important in achieving a larger goal. This is the foundation of continuous improvement. If you wait for perfection you may enter analysis paralysis (analyzing and not doing).

The supermarket concept is based on the supermarket in your community. You go to the store, pick what you need when you need it, and deliver it to your home. You only spend money on what you need. After all, things

FIGURE 9.8
A supermarket.

spoil, and that is a waste of money right? But parts don't spoil you say. Don't they?

Parts do have a shelf life. Gaskets get dry rotted, dust and dirt collect on parts, and things wear out or are not the latest version. Besides, why would you spend money on it if you don't need it right away? The supermarket will help you understand this better for future inventory improvements.

First in first out (FIFO) is a way of using the parts first that you receive first. Again, the concept comes from your neighborhood supermarket (Figure 9.9). For instance, the milk isle at your friendly neighborhood supermarket holds dairy products on gravity fed racks, with the oldest milk on the shelf first at the point of use of the shopper. This ensures that the consumer is practicing FIFO. Your parts supermarket needs to be set up the same way.

The two-bin system has two bins of parts (quantities together equaling one day only on the line) set up for FIFO in your supermarket. You will need to understand the daily demand for all parts on the line plus a small buffer. Set them up on a two-bin system and this will help determine the size of your supermarket.

You must also take into account room for the material handler picking parts and the tugger or cart (to pick up full kits, return empty kits, and return to the line). The tugger or cart delivery and return is called a milk run (Figure 9.10). The term *milk run* comes from the days when a milk man delivered milk right to your door. You placed your empty milk containers

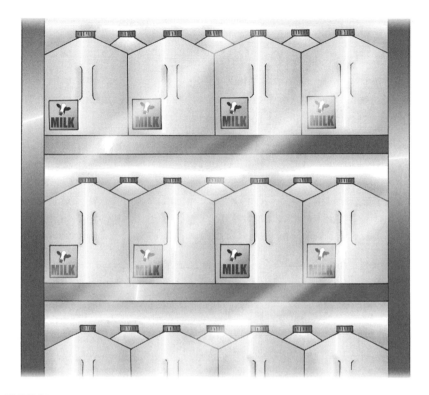

FIGURE 9.9
The FIFO concept with milk in a supermarket.

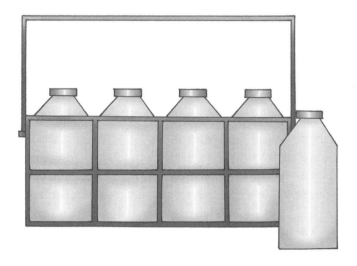

FIGURE 9.10
The origin of the term *milk run*.

outside of your door and the milk man would pick up the empty containers and leave you full containers of milk.

The same principle applies in a two-bin system; the empty container is a signal that you need a refill. This is going to happen in your supermarket, which is resupplied by the warehouse, as the material handler brings back empty kits from the line for the supermarket to fill.

NOTE: Do not place your supermarket next to your production line. This is very important! As you begin kitting, you will find BOM issues. Operators want to make products on time, so they will rob the supermarket for the part they need. However, this will not help you resolve your BOM issues. Not to mention the mess that it will make of your market when a lot of hands are in it.

No one should be allowed in the supermarket except for designated line and supermarket leaders. We will discuss how to correct your BOM issues and kitting issues later.

Typically pick lists to the warehouse are based on material requirements planning (MRP) systems and printouts. Remember that you are running a two-bin system in the supermarket as a one-day supply to the production line.

Processes for replenishing from the warehouse to the supermarket need to be defined. Delivery from the warehouse, replenishing the supermarket, and pick lists should follow a similar process as the supermarket to production cell replenishment.

PICK LISTS

Pick Lists and *Heijunka*

Pick lists are developed from your BOM. They are a grocery list of what you need to select from your supermarket to deliver to your production line. The pick list needs to be separated by the workstation's content. The parts need to be in the kits delivered to that workstation or line.

For one-unit kits this is relatively easy. The pick list is the BOM in sequence of the build according to the standard work. But for multistation kits this is a complex process. As a unit moves down the line, it passes from one station to the next and different parts are installed at each workstation. If a tugger delivers all the parts for one unit at a time,

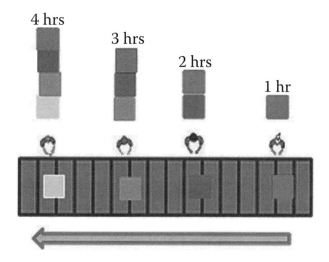

FIGURE 9.11
Demonstrates parts deliveries without the use of a heijunka. The last station contains piles of parts.

the last station on the line will always have a pile of parts because the unit has not moved down the line to the last station yet. The parts build up on the line and defeat the purpose on one-hour line-side inventory and kitting (Figure 9.11).

To resolve this issue a *heijunka* is needed. *Heijunka* is a Japanese term meaning to "level the load." It is a technique used to level production to meet demand. You used a form of level loading in Chapter 2 when you determined the daily demand for takt time. For kitting, the *heijunka* is used to bring each station the parts it needs only when it needs them, that is, just in time (JIT). In Figure 9.12 you see that by using a *heijunka*, each workstation can receive parts for the unit that has moved down the line into their work cell. Note that the tugger actually contains parts for several different units on the line. In fact, it will hold parts for as many units as you have workstations. This is why one-unit kitting is much easier. For example, a pick list for a tugger in this situation will contain different parts for different units for different workstations (Figure 9.13).

The material picker should have a sign-off section of the pick list to ensure that all parts are picked and so you can identify the need for more training with employees. Advanced facilities use a barcode system and scan their pick lists. In this system material is moved virtually to a line-side location and then at the end of the line a scan virtually places and checks your materials into a finished goods location so that all inventory

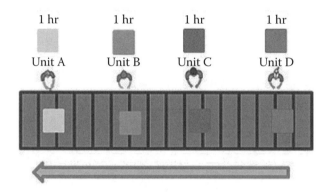

FIGURE 9.12
A balanced working cell with JIT levels of inventory.

is accurate at all times in your system. In addition, some facilities have installed picking light systems to aid in the picking process and ensure accuracy in picking parts.

To start kitting for the first time you must empty the line and start from scratch (Figure 9.14). The first time the tugger will arrive with only parts for workstation number 1 (unit A). The next tugger delivery will arrive with parts for workstation number 2 (unit A) and workstation number 1 (unit B), and so on, and so on, until the line is full. From this point forward the line will run according to *heijunka*. Daily status for every unit is also very easy to determine, as you will know from the moment the unit is scheduled where it is in your facility because you will know the exact time of every workstation.

MATERIAL HANDLING TAKT TIMES

Material Handling Is Now on a Takt Time

Employees will need to be assigned to pick parts from the supermarket. This doesn't mean that you have to hire a lot of people. You have eliminated so much waste from your line that you will free up manpower that can become supermarket pickers. You just redeploy employees to that area. After all, everyone in the work cell has been gathering parts and adding cycle time to your product, which increases lead time and deteriorates productivity.

Line 1   **Pick List**

Station 1
Unit A

Part #	Picked	Notes
24598463		
34598759		
21569872		
24598752		

Station 2
Unit B

Part #	Picked	Notes
21459875		
23654128		
21547896		

Station 3
Unit C

Part #	Picked	Notes
25489631		
59874598		
23665447		
22456987		
12366548		

Station 4
Unit D

Part #	Picked	Notes
21458777		
23654128		
14566321		

Material Supermarket Picker Signature: _____

                                                    Date: _____

FIGURE 9.13
A pick list.

Guess who else needs a yamazumi chart and standard work? You guessed it, all of the material handling. To determine the takt time for the tugger driver, consider that they must arrive at the line before each workstation runs out of parts.

If you are kitting multistation kits, the tugger driver must deliver to all stations on your line before they run out of parts in the kit they are

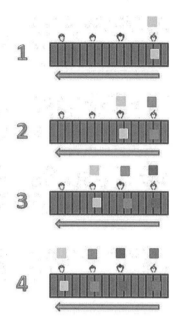

FIGURE 9.14
Wetting a line for the first time when you begin kitting.

currently working out of and pick up empties to take back to the super-market for refill. This requires a lot of planning if your line contains work-stations on both sides of your production line.

The material handler will move the tugger up the line, return to the supermarket, drop off empties and pick up full kits. All of this must be completed in time for the kits to arrive before the workstation runs out of parts. This magic number is the takt time. The placement of your super-market may be a factor here. Just remember, do not put the supermarket close to the line.

You will need to simulate the time for the tugger's arrival to the line, material delivery to the workstations, pickup of the empty containers, and the tugger's return to the supermarket. This time is needed to determine the tugger driver's cycle time. It can be simulated by riding around on a bicycle or forklift and imitating their work content.

The supermarket pickers work content needs to be simulated as well to determine their cycle time. Their takt time is based on gathering an empty kit, picking the parts for the kit, signing off on the pick list, and placing full kits in the tugger pickup area. All of this needs to take place before the tugger driver returns.

Since picking parts is also on a takt time, it is important to reduce the waste as much as possible by arranging the parts in the sequence in which they are to be picked according to the workstations. Wow, everyone is now running more efficiently and to the rhythm of the factory. (Remember to set work content no greater than 95% and no less than 85% of the takt time. Adding more or less time is adding waste.)

You have now determined workstation standard work, kit size, workstation design, material delivery design, line layout, supermarket design, tugger routes, and material delivery standard work. So what happens when things go wrong as you start this new process?

REFINING THE PROCESS

Correcting Mistakes and Anticipating the Crash Period

Major improvements contain a crash period. This is caused by the learning curve and adjusting to the new process (Figure 9.15). You need to expect it and prepare your team members for it. It is inevitable. In the future, when you work through the issues, you will perform at a higher level than you were before the changes.

When it comes time to implement the new processes you will need some intense training for your team members. To assist with this, give each team member a copy of their yamazumi charts and standard work one week in advance to study and ask any questions they may have. When you remove all parts from the line (and stock them in the supermarket) you will install new workstations. At this time, place large Xs on the floor with workstation numbers at the locations where each team member will stand when performing his or her standard work. Stand with each team member on the line and simulate the member's new work content to ease him or her into the new process.

FIGURE 9.15
The progression through Lean initiatives and the crash period.

You will need leadership support to be stationed line side and in the supermarket for at least a month after the initial implementation phase. The line-side support is stationed to assist the team members with parts shortages and excessive parts. Shortages of parts need to be verified with the BOMs and the line-side engineer to correct issues right away to ensure that the next pick list is accurate. If the BOMs are accurate and parts are missed, the supermarket leadership will assist the picker who signed the pick list in correcting the issues and providing support.

A process needs to be developed for corrections to excessive and missing parts from the supermarket. The supermarket support should be responsible for getting parts and making corrections to pick lists or picking errors. All of the errors should be documented and tracked for problem solving. These data should be used in daily meetings with the team. Dedication to the process will produce great rewards when issues are resolved.

In the near future this process will sustain itself and your old issues (downtime, missing parts, incorrect parts, hunting for tools, quality issues, loss of productivity, and lead time) will soon be a thing of the past.

NOTE: Start this process with a pilot cell and when all issues are a thing of the past, move to the next work cell. A pilot cell can be completed in one month with dedication and determination.

10

High Mix and Complexity

So that all sounds wonderful, but you have a business where no two products or services are the same, or your mix is very high. Typically products or services with total operator cycle times that have a 30% difference should not be performed in the same value stream (on the same line). This causes bottlenecks and constraints that cause delays in meeting customer demand. Remember the pipe?

For job shop areas, identify the units that are 60% to 80% of your build and balance your work to those services or products. You don't want to account for the longest cycle time product if you rarely run that unit. If you do, you are inserting the waste of waiting into your system when you perform easier tasks. It is also possible that you need to take an average and balance work on the average cycle time. For these situations, one-unit kitting is typically used.

These Lean principles and concepts are valid for any type of business you are in. You must not stray from Lean principles, but you will have to make it specific for you and what you do. Productivity is a balancing act. Once you have removed the majority of your unnecessary non-value-added activities, you will be surprised at your gains in quality, on-time delivery, lead time, and throughput.

11

Conclusion

Now you have

- Completed time studies
- Created yamazumi charts
- Created spaghetti charts
- Created standard work combination sheets
- Identified wastes
- Reduced waste
- Created standard work
- Established 5S
- Rebalanced the line
- Developed kit sizes
- Determined material delivery methods
- Created workstations
- Created defined tugger routes
- Created a supermarket
- Created pick lists based in *heijunka*
- Established standard work for material handlers
- Established warehouse replenishment

You have made gains in

- Productivity
- On-time delivery
- Lead time
- Quality
- Costs

It seems like a lot of work, but what things in life that are worth having aren't? The next step is to evaluate the supermarket and warehouse after about a year and push the supermarket back to the warehouse and reduce inventory from your suppliers. To do this you will need to evaluate smaller, more frequent delivery methods.

Eliminating waste through line balance and JIT kitting is a very exciting journey that will mature your facility to world-class standards. It will be very rewarding to see your facility mature into a well-oiled machine and to be a part of the process of change for the future.

I hope you enjoy your Lean journey and strive for success. Remember that it does not happen overnight. It is a relentless pursuit of *continuous improvement*, and oh-so-rewarding.

Index